Faces of the Future

The struggle for the next generation in Paktika Province, Afghanistan

Photos by
Brad Israel

Photography and text by Brad Israel
Designed and edited by Richard McClintock
Printed by Smitty's Printing Company, Columbia, SC

ISBN 978-0-9895303-0-9

Brad Israel was born in Mobile, Alabama, and graduated from Hampden-Sydney College in 2004. He joined the military in the spring of 2005, receiving his commission through the Officer Candidate School (OCS). Brad attended the Infantry Officer Basic Course, Ranger School, and Airborne School before being assigned to the 1st BN, 503rd Infantry Regiment, 173rd Airborne Brigade Combat Team in Vicenza, Italy. While in Italy, Brad served as a Rifle Platoon Leader and trained with his men for a fifteen-month deployment to Afghanistan in support of Operation Enduring Freedom. In Afghanistan Brad served as a Heavy Weapons Platoon Leader and Scout Reconnaissance Platoon Leader. His unit returned to Italy in July 2008, where he remained until June of 2009. Brad has since deployed back to Afghanistan, serving as a Detachment Commander in 7th Special Forces Group.

Faces of the Future

Fran,

Thank you for all the incredible sacrifices you & your family have made over the years. Please accept my sincere gratitude and utmost appreciation. I hope these photos provide a different perspective to some of the many things we do in Afghanistan. Merry Christmas!

2013

"Know the people, the topography, economy, history, religion, and culture. Know every village, road, field, population group, tribal leader, and ancient grievance. Your task is to become the world expert on your district."

DAVID KILCULLEN, "TWENTY-EIGHT ARTICLES"

Foreword

This is a story without a happy ending, so far. Its conclusion hangs undecided in midair. It tells of one man's efforts to bring security and hope to a people living in fear under harsh conditions reminiscent of the Middle Ages.

As Captain Brad Israel, H-SC '04, was to learn from multiple tours in the Afghanistan conflict, the face of War has changed markedly since the end of World War II over a half century ago. Revolutionary scientific breakthroughs and correspondingly dramatic advances in technology have so exponentially increased the destructive power of modern weaponry that co-belligerents have retreated—in a sense—to the lower end of the conflict spectrum where it is safer to fight. There they find that War is much more than a military venture; the political, sociological, economic, and psychological dimensions are now at least as important as—and at times more so than—the purely military.

In wars of the 21st Century, the primary objective in ground combat is seldom to seize and hold a critical terrain feature, such as a vital road junction or an important hill—a "goose egg on the map." The goal for which contending forces now strive is the heart and mind of a human being, the poor unfortunate caught in the middle, often in a deadly crossfire. The Chinese Communist revolutionary leader, Mao tse-tung, had it right when he wrote "...the people are the water and the guerrilla-insurgent the fish...The fish swims in the water." And more often than not, the ultimate outcome will be decided in favor of the contender who succeeds in gaining (or in some instances, coercing) the support of the people.

That is the message of the photographs contained in this book.

Samuel Vaughan Wilson
Lieutenant General, US Army-Retired
President Emeritus, Hampden-Sydney College

The author with Afghan children, the faces of the future

Faces of the Future: The Pashtuns of Paktika

The purpose of this book is to bring to light the remarkable unity and sense of accomplishment of a group of young men, a support base in the United States, and the Paktika people. This diverse triangle of cultures symbolizes the unique and fascinating ability to work tirelessly and together towards common goals, specifically, at its core, helping the next generation.

Education will always promote opportunity and opportunity will provide the potential for a better quality of life. This better quality of life can bring hope to a region of the world that desperately needs it. Perhaps the next generation of young Afghans of awesome promise and potential can bring economic growth and stability back to their sub-tribe or village. The select pocket of Pashtuns, with whom we worked and were able to influence positively, understand the significance of this righteous pursuit. Just as we want our children to have a better life and more opportunity than we did, so do the Afghan parents for their children.

One of many stories about our support base back home: my Mom was active in aiding us from miles and miles away, and once she sent us a portable color printer. We could take it to the villages, snap photos of the children and their families, and within minutes give them these photographs. They were astonished; they had never seen anything like this and were absolutely thrilled to have them. It was an instant rapport-building tool.

Near the end of my deployment, as I staged my gear on the helicopter landing zone (HLZ) at FOB Curry in Gomal, mysteriously a handful of village elders began to convene at the front gate with the sub-governor. I had not told them when exactly I was leaving, for operational security concerns, but somehow they already knew.

I waved them in after a search and met them inside a guard shack on the HLZ. Scooter and Tyson, two of the finest interpreters I had seen and worked with in Afghanistan, brought in bread and chai (tea). We sat around and broke bread and drank chai one last time together. The village elders presented me with a tailored Afghan Kamiz Shalwar (a matching top and trouser). We reflected on the months spent getting to know one another and all that we had accomplished.

I told them how important it was to stay the course, no matter what American unit replaced us. These elders held the future in their hands and it was imperative that they promoted the goals of what they envisioned as the way ahead in Gomal. The last topic of conversation was education and the security for their children. We agreed this was the most powerful project that needed to be pushed the hardest. They pledged to keep the schools functional, whatever the sacrifice or burden on the villages and teachers. I let them know I would pray for their strength, success, and resolve with all these challenges that lie ahead. The elders prayed for my men and me and for the happiness and welfare of our families.

As I got the call that the helicopter was three minutes out, we hugged and said our final good-byes. The elders began to wipe away tears, and I had a knot in my throat and could not speak anymore. It was difficult because I do not know if I will ever see the likes of any of those noble Afghan giants again. However, it was also one of the most rewarding moments of the entire deployment. My men and I had successfully touched these men's and this tribe's hearts; we appealed to them on an emotional level and brought humanity to the forefront of our mission. As I stepped off towards the Chinook helicopter, Malam Rasa, the key leader of the Gomal district, grabbed my arm and said, "Inshallah, we shall meet again." I looked back at him and all I could get out was "Inshallah" with a smile. Inshallah means God willing or God's will, and the Pashtuns live their life by this word, everything they do or fail to do is from God's will.

Acknowledgements:

First and foremost a special thank-you goes to all the men who allowed me to serve as their platoon leader in combat. There are few things more incredible in life than being responsible for a powerful airborne infantry platoon, be it light, heavy, or the Battalion Scout Recon Platoon. It is a magnificent force with superior capabilities. I was fortunate enough to be with all three. These men were nothing shy of a courageous group of self-sacrificing patriots who wear many different hats, from that of a warrior to that of a teacher or distributor of humanitarian aid. The roots and strength of our military rely on the heart and conviction of these young infantrymen.

An impossible debt of gratitude goes to those who support these heroes. My mom, dad, brother, sister, as well as extended family and friends, provided me with a steadfast backbone of strength and support throughout my military career, specifically during this deployment. Thanks to mom's tireless work in my community back home, she raised the awareness of the plight of soldiers and Afghans and brought the reality of war to our hometown. My community in Mobile, Alabama, raised thousands of dollars to support my men and properly field them with hiking boots to fight in the rugged, unforgiving Afghan terrain as well as with lighter, more extreme-weather British sleeping bags to survive the harsh winters while out on extended operations. My grandmothers, aunts, uncles, cousins, and friends seemed to be there every step with me; I love them all dearly and am very grateful for their love and care of my men and me.

Next, to all the families and individuals who adopted my platoons, people from New York to Florida and California, Texas, Georgia, Michigan, and Virginia, even from Israel and England, who sent my men supplies. This was by far the most humbling event I witnessed. The perplexed look on these soldier's faces when packages of socks, hygiene products, magazines, games, linens, and food would come to them was priceless. They would ask me why people who did not know them would send these expensive items. I told the guys they did it because these people cared about them. What these individual supporters and families may not know is that they were not only equipping these men with essential items and some luxuries, they were also teaching an invaluable life lesson about caring for one another: "Love your neighbor as I have loved you." This was a powerful example of the Lord actively at work in our lives. So if any soldier ever forgets why we fight, I would be more than happy to share my story. There will always be people back home, citizens from every state, strangers from coast to coast, who love and honor you, the soldier, even though you do not know them and they do not know you. The people of the United States of America are our nation's most powerful attribute. For the most part, they embody what makes America great.

I also must acknowledge my Afghan counterparts, who put their lives and the lives of their families on the line every single day. They provided my men and me a wall of security that the enemy could not penetrate. On multiple occasions, they won the battle for us before it even started. I praise these heroes for their courage, sacrifice, hope, and ability to survive. If one ever wants a glimpse of how life was lived in the biblical days, visit a remote Afghan tribe with no access to roads, who have been self-sustaining for generations. Always be grateful for what we have back home.

My mom, dad, and wife Anna have stayed on me to put this together. While it may only scratch the surface with some of our experiences during Operation Enduring Freedom, I only hope it opens the eyes to those who do not believe good and noble things can be achieved in this war. Asking a young man who just survived an ambush and has lost his friends on the battlefield to sit down, lay his weapon to his side, and talk about tribal infrastructure and future development of their village, is not something you can ever train or prepare these men for. Be thankful these patriots exist and are willing to do this so that others do not have to.

To all those who supported us, thank you for believing and trusting us to represent you abroad. We all proudly wear the flag to honor those who have laid down their lives and to those back home who are with us every step of the way.

God Bless and thank you for the help. Your number will be called time and time again, and thanks for stepping up when you can.

I only hope that these short excerpts and photos reveal the ability to positively influence and assist others, regardless of the profession or career path one pursues. We must always continue to support one another and our way of life.

Thank you, General Sam, for your foreword, but more importantly for your mentorship over the years. Those who know your story will always share it with others; you have accomplished more in life than I could ever hope to. Thank you for teaching the next generation; your legacy will live on. A special thanks to my brother Will, who time and time again is called upon to proof my work; I appreciate your making me appear more intelligent than I actually am.

Finally, it is because of my wife Anna that I did not give up on doing this, although it is now four-plus years removed. Thank you for believing in me, for keeping me on track, for always being such an incredible wife and best friend, and for blessing us with our son Lyon. We are what we are because of you; I love you.

Faces of the Future

Eyes of the Future

A Desire to Learn in Gomal

This was a funny excursion. The school principal was having an extremely difficult time controlling and disciplining these kids, outside a children's school in Gomal that a Non-Governmental Organization built in 2006 and then was burned down by the Taliban not even six months after opening. The teachers and children still use the building structure to teach and learn; Afghans always adapt and make do with what they have.

The school bags with cats on them (on the next page) were pre-made school supply kits that an incredibly smart woman back in the United States put together for us. She was a steadfast supporter in helping us reach out to the local populace, and certainly did her homework. The significance of the cat, as the legend goes, is that Mohammad liked cats because one saved him from being struck by a poisonous snake. He petted the cat and thanked her for saving him. Because his hand stroked the cat's back, cats never fall on their backs. Cats are therefore respected and acknowledged in the Islamic culture.

This was a police checkpoint that we came to visit. I laughed because, as normal in a village, most of the police guards were either in the village buying something or sleeping. When this group of boys heard me, they all popped their heads up and said they were the local police for the day.

Hoodie

A new jacket

We gave the water gun to this kid because he was courageous enough to take us to an area laced with improvised explosive devices. He pointed out where each bomb was buried, and we were able to safely reduce, recover, and destroy them in a safe area.

Playful Opinion

This was a local policeman's son, who looked out for us and brought us fresh vegetables when they had them. He was friendly, curious, and disciplined for an Afghan boy.

Reaching for Knowledge

Blossoming relations

Father and son sharing breakfast

NIDO

Group Instinct

Friendly Face

The Gift of Giving: An Afghan Christmas

This was on Christmas day and was considered our cross-culture sharing of gifts in order not to impose Christianity on our Muslim friends. We explained our holiday and offered those presents to celebrate with us. Over 150 kids from as far as 25 miles arrived early on the 25th of December, most traveling through the night with their fathers, by foot, camel, mule, motorcycle, and truck. The smiles on their faces will remain one of the most positive memories I have from the deployment; every one of these gifts was sent from the incredible supporters we have back in the United States. My mother, friends in our community, and others across the states made "Rag Dolls 2 Love" for us to hand out. These dolls were recognized by the women back home who made them and gave positive physical feedback on their efforts and commitment. This was a perfect example of connecting those supporters back home to the real world events on the ground.

We also distributed quilts, gloves, used clothes, toys, and hygiene products that they sent from home. The ability to connect those back home to what we were doing on the ground was a humbling experience, and I am very fortunate to have been a part of it. I would often tell the Afghan children that the clothes they were receiving were from a girl her age in the United States who cares about her and her future. Being able to come together, regardless of our backgrounds, is a truly remarkable example of what makes our nation so great. The impact of our many supporters was felt thousands of miles away.

A warm appreciation

Sporting a New Coat

A long way to Disney: Afghan Mickey fan

US ARMY

Afghan hearts

Positive feedback

Afghan National Army (ANA): Eyes on Pakistan

Afghan National Army (ANA): Smoke break on the border

Sunrise breakfast boy

The Importance of Chai

Bandar was one of the friendliest villages within walking distance of our outpost. This little guy joined us for a sunrise breakfast outside of the mosque. The elder, our Pashtun Warrior, was a phenomenal man; we gave him medical attention our first week in the Gomal district, and ever since, he pledged his commitment to us. They served us hard-boiled eggs that morning when all we had been eating for weeks were MREs (meals ready to eat). Those may have been the best eggs I have ever had! These villages became our eyes and ears to the east; no one would be able to attack us without an early warning from these courageous few.

It was at this village, months before, that I realized the importance of drinking chai with the local elders. The Afghans absolutely love their chai, and they pride themselves on being hospitable. It was customary to be offered chai when meeting locals in a village, and it was considered rude not to accept their warm invitation. This was an enormous part of their culture and was their way of conducting business. There have been countless discussions, recommendations, and resolutions since the beginning of the war that have all come to pass over a cup or two of chai. This was a significant act in order to get a "foot in the door" with some of these tribesmen. We purchased many boxes of chai on the local economy to give to the elders as a gift when we came to talk and learn.

Egg breakfast

Spera and Universal Childhood

The photos on the following pages are from one of the worst areas in Afghanistan, an area rich in history of fierce fighting dating back to the Soviet invasion. There is an extensive network of caves, hideouts, and enemy trails running all along the porous border. We stayed in this area for over 30 days, conducting operations and working with the locals. Every day in this area was dangerous. It was very hard to tell whom to trust and who was collecting intelligence on our location. On September 11, 2007, one of our Rifle Companies came into contact with over twenty Al Qaeda fighters who were using this area as a staging camp for operations and facilitating border crossings. When we handed out some of the clothes and humanitarian aid, these tribal elders did not know what to think. They ended up being happy with our presence, but after we left, this was one of the first places to deteriorate. It was handed over to the ANA (Afghan National Army), and our replacement unit was quickly called back to that area after the outpost was overrun and the ANA either fled or were killed.

The kids in this area are being bred to fight. One can sense it in their mannerisms and attitude. They are not representative of children in other areas that we visited. This area will continue to remain the "Wild West" until the border is recognized and properly manned, more access roads for commerce are established, and children are educated properly so that they can learn a means of living that does not revolve around weapons and radical madrassas.

I quickly learned how common kids are all around the world. At a young age, we are all very similar, but as we grow up, these fundamental years of development encourage and influence who we become. This is our foundation, and we must identify what it is in this development process that goes wrong in the Afghan upbringing. These photos capture how for the Afghan kids find out that there is a human side to the soldier behind the gear. It allows us to break down the wall, poke fun at ourselves and each other, and laugh together.

Afghan leader

Wise Youth

Shura in Spera

The Slingshot Gang

A good friend of mine in Atlanta sent me dozens of clown noses and hats that are normally handed out to kids at the circus. She thought it would be a great idea to distribute some to the Afghan kids. She was right; they thought it was absolutely hilarious.

Making Time for Laughter

The confident and courageous clowns of Bermel

Scarf Humor

Fashionable in Spera

Often the humanitarian aid packages sent out to the remote sites were completely random. This one happened to have colorful women's clothing mixed in with bags of rice and beans. The locals were as puzzled as we were about where these came from. Thankfully, they laughed and took them to distribute back amongst their villages.

Beanie Baby bandit

The backpacker

Gifts for the family

Girls take interest

The Afghan squat

Sunglasses Princess

As we passed through a wadi in the Sarobi district, our convoy passed this young girl walking along the rocks. I had some candy and toys in a bag, and I found this pair of sunglasses that some kind neighbors from back home sent to me.

Sunglasses princess

Family portrait

Malam and Daughter

The plateau we occupied was the very ground that Haji Malam used to defend. Five years ago he and 20 others fought off over 100-plus Taliban in one engagement; he still does not know how he made it out alive. " Inshallah" – "God willing" he tells me proudly. Haji Malam earned the prestigious title of Haji after taking his hajj to Mecca in the fall of 2007. He brought me a beautiful fountain pen upon his return. This was a powerful token of appreciation; it proved to me that he did like having us around. His support was our security. It is amazing how much power one man can have—such a person really can influence the successes or failures in an area.

Flying the Flag in Charbarran

These kids lined up in the field as we handed out Afghan flags and humanitarian aid. Our Company had a difficult time in this district. These locals were being coerced by Taliban and we could feel the tension and nervousness of the tribesmen whenever we visited. It was obvious they were always being watched during our meetings. The children loved all the gifts, though, and in the end that was one of the most important building blocks for our relationships in Afghanistan. The small victories come through focusing on the next generation, so that they can grow up with the opportunity for a better life.

National pride

Showing off our new flags

Gomal Hygiene Class

Whenever we had extra toothpaste, toothbrushes, and hygiene items, we gave the locals classes on sanitation and basic hygiene. Every time we visited this village the kids would run out yelling and holding their toothbrushes high into the air. After the lack of clean water, the near total absence of hygiene in third-world countries is one of the most significant causes of disease and sickness.

Tyson, on my left, as well as my other interpreter Scooter, were truly gifted in expressing my body language and tone. In this photo, we were joking with the kids, keeping them at ease, and having a good time. At times, if I were frustrated with the tribal elders or needed to convey a serious message, the interpreters would almost mimic the tone in my voice, my posture, and body language perfectly. One can only imagine what all gets lost in translation, but when an interpreter can exercise these traits, he or she is definitely very special. We were able to achieve much of our success and develop so many strong relationships because of the heroic Afghan interpreters who went out with us every single day.

Practicing before the first brush

Boy on Camel

We passed this family of Kuchis (nomads) in Gomal while headed on a patrol towards the border. The Kuchis live their lives moving from one region to another in search of work, hospitable tribes, and safety. Some may risk their lives and the lives of their families to help out the coalition effort. Because of the way they roam, they tend to stumble across weapon caches, enemy observation posts, and rocket launching sites. These Kuchis know there is opportunity for work if they want to assist the efforts to rid the area of enemy. One Kuchi brought us to an enormous weapon cache where there were armor-piercing rounds, rocket-propelled grenades, mortars, and a brand new satellite phone with number prefixes to all over the world.

Little Kuchi's trek to new land, Bandar

Magical Lamb Jerky

This was my police chief; he found his way into the top role by being an extremely charismatic and tactically savvy soldier. The former police chief was corrupt and worked with Taliban in Gomal, so once we relieved him, Tofan was the natural choice. Tofan became one of my good friends and helped create a security bubble in Gomal, always trying to protect my men and me.

During Ramadan, after a chance contact and successful offensive engagement, I brought the enemy weapons back to the district center to give to the police force. He yelled at me for not waking him up when we left and said it was his job to protect us and keep the area safe, not mine (even though the Pushtuns stay up all night during Ramadan eating and conversing, and they fast during the day).

Here he is teaching me how to savor all the meat off a goat. Every time a goat is slaughtered, these tribal men will go through their own prayer rituals to thank Allah for their feast.

Police chief's magical lamb jerky

Little Boy with Marker

This little boy was proud of his new marker. He was even more shocked when he got new winter clothes and other school supplies. About 20 minutes west of his village, we found a large cache of weapons and rockets. One of the tribesmen from this village led us to the site a few weeks following this aid distribution.

Making his mark

New gloves

Surviving the Winter

These brothers were out searching for wood in the Charbarran district as the snow started to come down. We stopped and gave them winter blankets, those purple and black gloves, and some food that we had, but unfortunately I did not have any wood. I have never seen more underdressed children than in Afghanistan in the winter. Children are barefoot in almost pajama-weight clothing and do not even seem to be affected by the severely low temperatures. I really began to push for children's used winter clothing from back home when I saw what harsh conditions they faced in the winter months.

The Bazaar Businessman at his Fruit stand

In the Bermel district, there was a fairly large bazaar. It amazed me to see all the brightly colored fruit, because after a few months the only noticeable color in this country was brown. These families worked incredibly hard to harvest any produce. Unfortunately, it was rare to see people purchase much fruit here due to the poverty.

A collection of colors

Outreach in Bandar

We were the very first Americans this family had ever seen. At first the man asked if we were Russians. When we gave them some humanitarian aid (HA), they began to weep in joy and could not fathom why we would drive all the way out to the middle of nowhere to bring gifts. “Bismallah” is all they could say, meaning “in the name of God.” We just happened to be surveying our battle space, and these few kalats (mud houses) sat off a major wadi (dry riverbed converted into a road) that appeared highly trafficable. Later, we found out this was one of the enemy’s main routes to get to southern provinces of Afghanistan.

Our Bandar brothers

The need for notebooks

King of the Village

The General's Seat

His father brought this young boy to a humanitarian aid distribution in the Gomal District of the Paktika Province. He was crying, so we made him "King of the Village," gave him a juice box, and set him up where he could see all the activity. We also gave him some new purple flip flops, because that was all we had left.

The Rock of the Family

This was a unique shot because these kids were following our vehicles in the Bermel district as we went to check the status on a village well. The girl definitely is the boss of her siblings and fellow tribe children; all the younger kids listened to her when they started to run wild. This holds true for most women in the Afghan culture; they maintain order and discipline within their families, yet are never given the credit. This photo was taken after I smiled and greeted her; smiling is contagious in every part of the world.

Afghan smiles

Boys are boys everywhere

The Future Governor of Paktika

This was the little bulldog. He held his ground and was not the least bit intimidated when we approached him. His brother was in the Afghan Army, and he was angry that he had not seen him in over a year. He told me to find his brother and tell him to come home so that he could have someone help him and his father in the fields.

The Bulldog

Discussing Politics

Warm hands and Warm Hearts

This was inside Bandar Checkpoint, later known as FOB Curry in honor of our 1SG who was killed in action on July 12, 2007.

This was one of my favorite days. Some of my local tribal leader friends and workers on our small outpost brought their children up for a visit (kind of like a "bring your kid to work day!"). However, the night before, we went on a mission and did not return until almost sunrise. As I lay down and fell into a deep sleep my interpreter, Scooter, woke me and said, all excited, "They are here, Sir!" At the time I was not as thrilled as he was. I said "Who?" and he replied, "The children, at least 20 of them!" I wanted to cry because I had just gotten warm in my sleeping bag and was expecting a solid five hours of sleep, but these kids were literally jumping up and down, they were so full of energy and excitement. I stumbled outside, still rubbing my eyes, and was happy to find such kind and hopeful kids, thankful for anything and everything. We played for a few hours, my interpreter showed them how to work some of the toys, we poked fun at their fathers, and they poked fun at Scooter and me, and they all left with bags bigger than they were, full of new toys and items for their family.

Pretty in pink

FARM ANIMAL PLAY SET

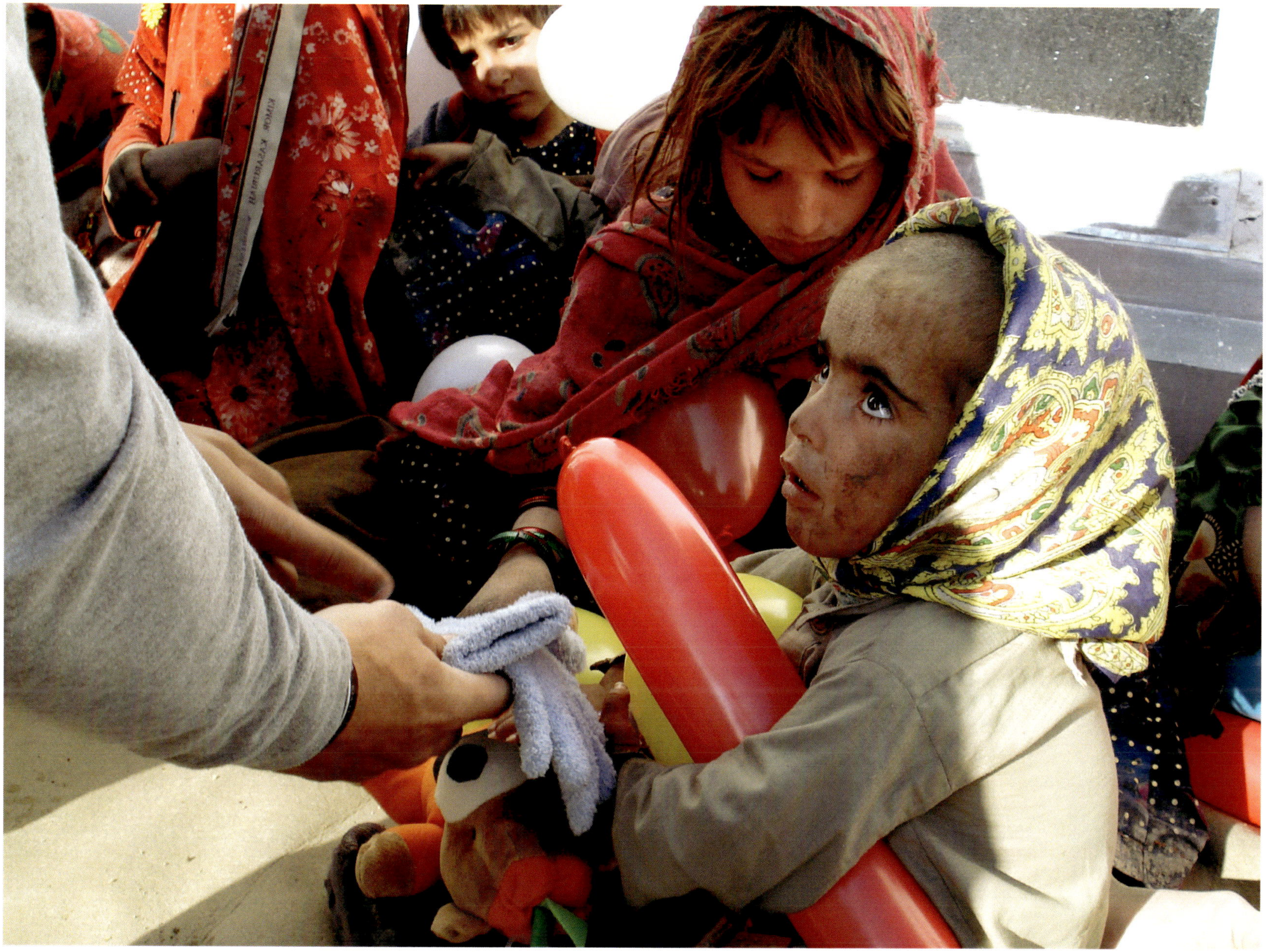

Confused but thankful

All-around excitement

Eyes telling a story

Covered in gifts

A handful

Adorable smiles

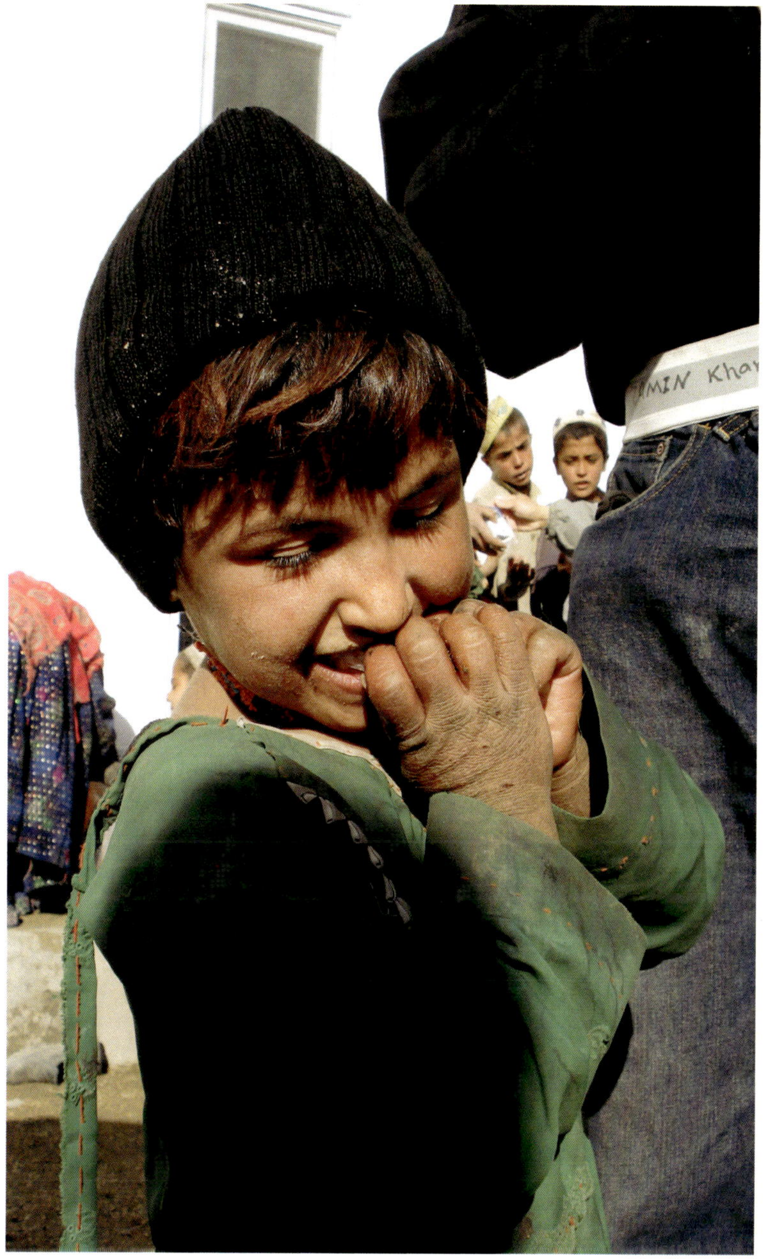

A wonderful day

A method of transport

She has talent

Connecting with the tribes

Father and daughter

More candy for this little one

Youth academic competition, Gomal District Center

I wanted to get the Afghan National Police (ANP) involved so that people in the area would start to become more receptive and open to share information with them. We held a "Youth Academic Competition" along with the school teachers and were able to talk about education in the area, and the potential for a new school and its safety and security. Tofan, the police chief (who was also my righthand man), helped organize the event. He highly exceeded my expectations by encouraging so many to attend. It was a morning for his men to shine: they quizzed the kids, gave out prizes, talked about the importance of education, and why they all wished they had had the opportunity to be literate and more educated. After this event, the police and our platoon began to receive more information on the insurgents than we could process; it was overwhelming and truly remarkable—we were achieving the effects that we wanted. Fellowship is significantly stronger than any other motivator.

Attentive and curious

The Sub-Governor taking care of his tribe's children

Transformers in Afghanistan

Spiderman's sidekick

No one leaves empty-handed

Sweet sisters

Tyson and the future of Spera

AFGHANISTAN

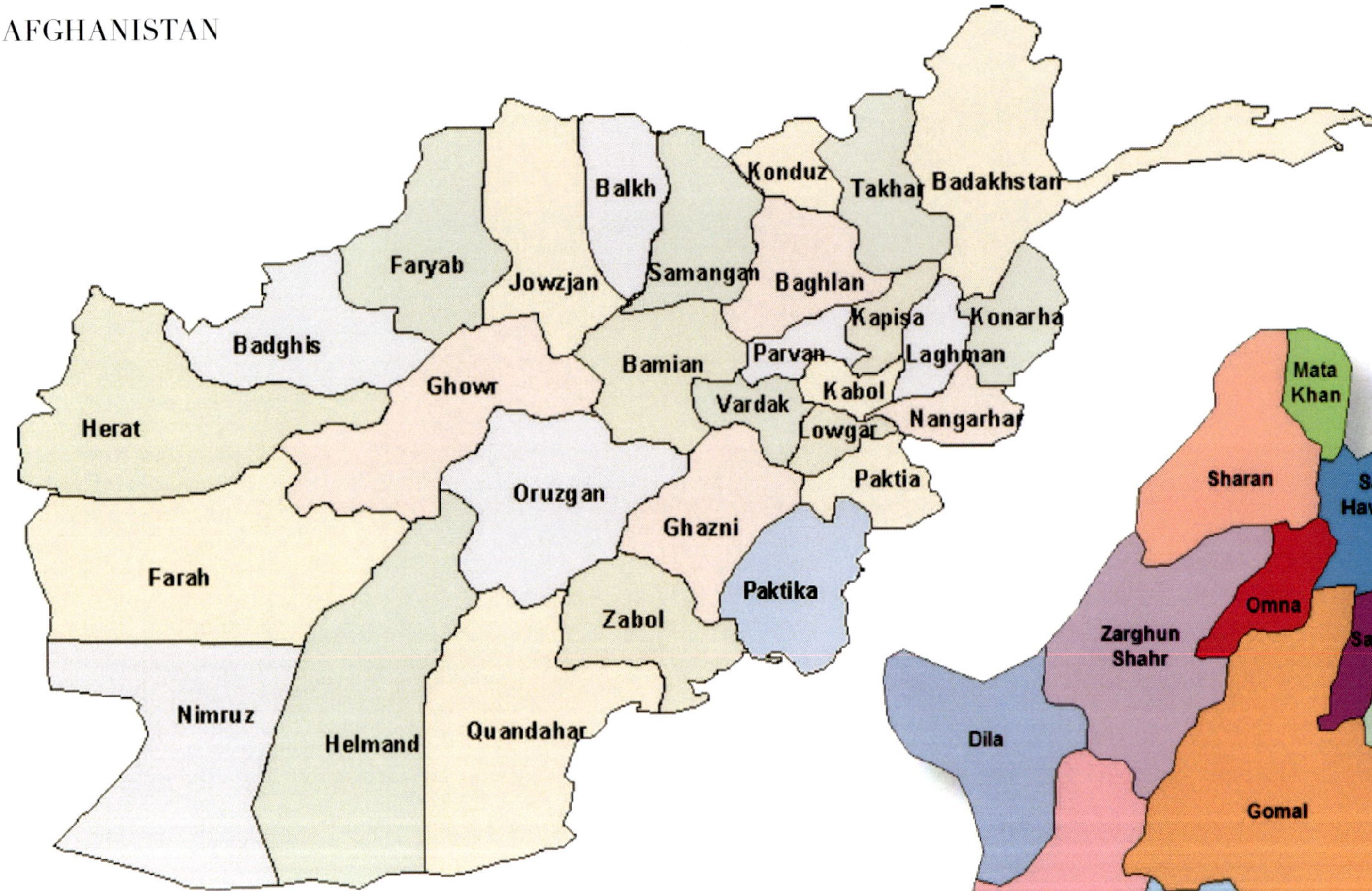

PAKTIKA PROVINCE

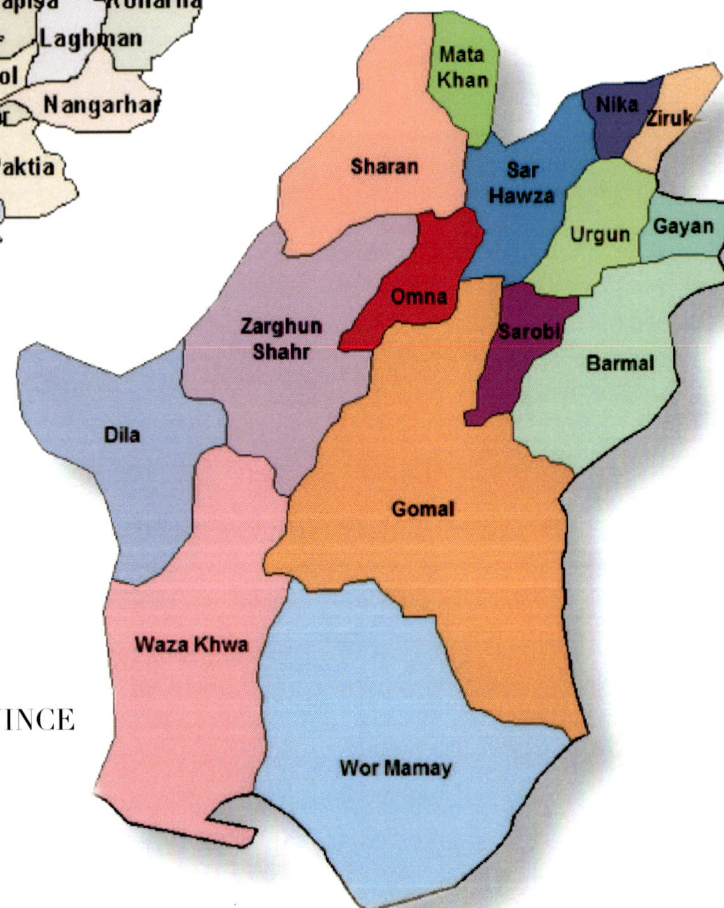

Maps courtesy of ReliefWeb.com